JA 21 '04	DATE DUE		
FE 23 '04			
NO 18 '04			
DE 08 '05			
AP 10 '07			
5-14-09 ILL			
OCT 06 '09			
JAN 05 '10			

BLAIRSVILLE SENIOR HIGH SCHOOL
BLAIRSVILLE, PENNA.

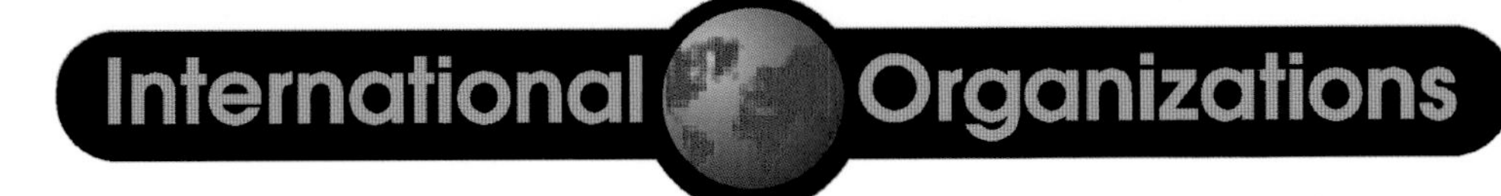

The ASPCA

Patricia Miller-Schroeder

WEIGL PUBLISHERS INC.

Published by Weigl Publishers Inc.
123 South Broad Street, Box 227
Mankato, MN 56002
USA

Web site: www.weigl.com

Library of Congress Cataloging-in-Publication Data

Miller-Schroeder, Patricia.
ASPCA / Pat-Miller Schroeder.
p. cm. -- (International organizations)
Summary: Presents the history of the ASPCA, discussing its origin, mission, goals, and achievements and providing some related human interest stories.
Includes bibliographical references and index.
ISBN 1-59036-024-9 (lib. bdg. : alk. paper)
1. American Society for the Prevention of Cruelty to Animals--Juvenile literature. 2. Animal welfare--United States--Juvenile literature. [1. American Society for the Prevention of Cruelty to Animals. 2. Animals--Treatment.] I. Title. II. Series.
HV4763 .M55 2002
636.08'32'06073--dc21

2002006566

Printed in Canada
1 2 3 4 5 6 7 8 9 0   06 05 04 03 02

Credits

Project Coordinator
Michael Lowry
Copy Editor
Tina Schwartzenberger
Photo Researcher
Gayle Murdoff
Design and Layout
Warren Clark
Bryan Pezzi

Photo Credits

Every reasonable effort has been made to trace ownership and to obtain permission to reprint copyright material. The publishers would be pleased to have any errors or omissions brought to their attention so that they may be corrected in subsequent printings.

Cover: Behling and Johnson Photography; **AFP/CORBIS/MAGMA:** page 9; **ASPCA Archives:** pages 7, 20, 21 top, 21 bottom, 22 top; **Behling and Johnson Photography:** page 15; **Pat Canova/MaXx Images:** page 13; **Comstock Images:** pages 8, 23; **Corbis Corporation:** page 25; **Corbis Sygma/MAGMA:** page 27; **DigitalVision Ltd.:** pages 5, 17; **Eunice Harris/MaXx Images:** page 11; **Lorraine Hill:** page 3; **Picturesof.net:** page 12; **Allen Russell/MaXx Images:** page 22 bottom; **Ted Streshinsky/CORBIS/MAGMA:** page 10.

Contents

What is the ASPCA?

The American Society for the Prevention of Cruelty to Animals (ASPCA) is an organization that helps protect the physical and mental well-being of animals. Some animals, both **domestic** and wild, are subject to violence and cruelty. The ASPCA was the first organization in North America formed to protect animals. Today, the ASPCA, along with the help of more than 680,000 members and donors, works to provide animals with better lives and brighter futures.

"Every step we take to increase the number of shelter pet adoptions means another dog's or cat's life will be saved."
ASPCA

The American Society for the Prevention of Cruelty to Animals has been involved in the humane treatment of animals for the past 136 years. In 1866, the ASPCA persuaded the government of New York State to pass an anti-cruelty law. It was the first anti-cruelty law in the United States to give an organization the right to enforce the law. This law allowed the ASPCA to investigate complaints of animal cruelty in the state of New York. The law also gave the ASPCA the power to arrest people who were being cruel to animals.

Most chickens are raised in overcrowded sheds. The chickens often have less than half a square foot of space. The ASPCA works to improve the living conditions of chickens.

The ASPCA employs fourteen Humane Law Enforcement Agents. The agents investigate more than 4,000 cases of animal abuse each year. The agents are trained at the New York Police Department's training academy.

Just the Facts

Founded: In April 1866, the American Society for the Prevention of Cruelty to Animals was formed in New York City. It was modeled after the Royal Society for the Prevention of Cruelty to Animals (RSPCA) in Great Britain.

Founder: The ASPCA was founded by Henry Bergh.

Mission: To promote **humane** principles, prevent cruelty, and alleviate fear, pain, and suffering in animals.

Scope of Work: With its headquarters in New York City, the ASPCA helps protect animals across the United States. The ASPCA provides humane education programs in elementary and high schools. It works to have federal, state, and local governments introduce laws to protect animals. It enforces anti-cruelty laws in New York State. The ASPCA operates a veterinary hospital, a mobile veterinary clinic, and an adoption program in New York City.

An Organization is Born

In the 1800s, many people did not think that animals had feelings. Horses, used to pull heavy carts and trolly cars, were often not properly fed or watered and suffered abuse from their owners. Stray dogs and cats were left to starve and were often injured or killed in the streets. Those people who were opposed to animal cruelty were powerless to stop the abuse.

"... a Society proposed to be founded for the Prevention of Cruelty to ... Animals."
The ASPCA Charter

The fight to put an end to animal cruelty was led by Henry Bergh. In 1865, he began to organize the ASPCA. Bergh modeled his new organization on the Royal Society for the Prevention of Cruelty to Animals (RSPCA) in Great Britain.

Bergh convinced a group of influential people to sign the charter for his new organization. Written by Bergh, the charter demanded better treatment for animals. In 1866, fifty people signed the The Declaration of the Rights of Animals. On April 10, 1866, the ASPCA was born and Henry Bergh was its first president. The ASPCA's first offices were in two small attic rooms.

By 1873, the ASPCA had become a model for humane societies that formed in Canada and twenty-five American states and territories. A giant step had been taken to protect animals in North America.

Quick Fact

One female cat and her offspring can produce up to 420,000 kittens in seven years. Every year, 3 to 4 million unwanted cats and kittens have to be put to sleep because homes cannot be found for them.

PROFILE

Henry Bergh

Henry Bergh was born into a wealthy shipbuilding family in New York City, in 1823. He received a good education and became a diplomat. In 1863, while working in Russia, he witnessed the terrible treatment of cart horses in St. Petersburg. When he returned to New York in 1864, Bergh saw the same cruelty to horses and other animals in the U.S. He was determined to stop the abuse of animals.

Bergh was a good speaker and he spoke publicly about the need to protect animals. He told people that it was important for humans to be kind and humane. While some people thought he was foolish, others agreed with him.

Standing 6 feet tall and sporting a large mustache, Bergh was an impressive and well-recognized figure. He earned the nickname "The Great Meddler" because of his constant efforts to stop animal abuse.

"Men will be just toward men when they are charitable toward animals."
Henry Bergh, Founder of the ASPCA

Henry Bergh worked tirelessly and seemed to be everywhere. He stopped dogfights and cockfights—where people bet on which animals would win. Many dogs and roosters were injured or killed in these deadly fights. Bergh saved as many fighting animals as he could. He stopped rat baiting events, where thousands of rats were collected from the city's garbage dumps and put into small rooms with a dog. People would bet on how long it would take for the dog to kill the rats or if the dog would survive.

Henry Bergh's first concern had been the treatment of horses. Under his guidance, the ASPCA pioneered the first horse ambulance in the world. He also created a special rescue sling to lift horses trapped in mud and water. Bergh was involved in setting up drinking troughs for horses around the city.

Bergh also cared for children. In 1874, he helped remove Mary Ellen, an abused girl, from her foster home. This led to the founding of the first Society for the Prevention of Cruelty to Children (SPCC).

Henry Bergh remained active in the fight to stop animal cruelty until his death in 1888. Today, the ASPCA is one of the largest and most active humane organizations in the world.

The Mission

The main goals of the ASPCA are to promote humane principles, prevent cruelty, and alleviate fear, pain, and suffering in animals. In order to fulfill its mission, the ASPCA is divided into different departments. Each of the departments has its own goals.

"Pets make a unique contribution to our psyche and our well-being, day in and day out."
Larry Hawk, President of the ASPCA

- The ASPCA Animal Science Department works toward making sure the latest scientific knowledge is used in helping animals.
- The ASPCA Government Affairs and Public Policy Department works to ensure that laws are passed to protect animals. The laws they have passed protect wild animals, **endangered species**, farm animals, pet animals, and animals in zoos, circuses, rodeos, pet stores, and laboratories.
- The ASPCA Humane Education Department provides educational programs and materials to teachers and students from elementary to high school.
- The ASPCA Public Information, Media Relations, and Advertising and Communications departments keep people informed. They give out news and information about the ASPCA's work and answer people's questions.

The ASPCA believes that education is key to preventing animal abuse. One of their programs is called "Learning to Care." It teaches children how to respect and show compassion to all animal species.

An estimated 70,000 kittens and puppies are born each day in the United States. Many of these animals are unwanted and are abandoned. The ASPCA has developed free spay and neuter programs to prevent unwanted births.

The ASPCA has many facilities that work to prevent animal cruelty.

- The ASPCA Humane Law Enforcement Office ensures anti-cruelty laws in New York State are enforced. The office works closely with police officers and gives advice to law enforcement agencies across North America.
- The Bergh Memorial Animal Hospital in New York City provides medical care to thousands of animals every year. It is one of the largest and best equipped veterinary hospitals in the U.S.
- The ASPCA Animal Poison Control Center, located in Urbana, Illinois, is the largest non-human poison center in the world.
- The ASPCA Community-based Animal Resources, Education, and Services (CARES) program operates two mobile veterinary clinics. The clinics provide free or low-cost vaccinations and **spay** or **neuter** clinics.
- The ASPCA Shelter Outreach Division helps animal shelters across the country.

Key Issues

The ASPCA is concerned about the many ways in which animals are exploited. The ASPCA has identified several key areas where the treatment of animals needs to be improved.

Entertainment

Animals have been used for thousands of years to entertain people. Sometimes the welfare of the animals is ignored as long as crowds of people pay to watch. Following are some of the types of entertainment that the ASPCA is concerned about:

"The ASPCA does not believe it is possible to train elephants, big cats, bears, and other wild animals to perform circus acts without abuse."
ASPCA

Circuses

Constantly traveling from place to place during a circus season can be stressful, especially in hot summer weather. The ASPCA is concerned that wild animals like bears, elephants, lions, and tigers are abused during training for circus acts.

Rodeos

Animals used in rodeo events can sometimes be injured or abused. This usually happens when animals have to compete in contests with other animals or with humans. Some events where animals are injured include: chuck wagon races, saddle bronc riding, wild horse racing, calf and steer roping, and steer wrestling.

Carnivals

Carnival events that may injure or cause stress to animals include: bear and alligator wrestling, greased pig contests, and donkey baseball.

In order to train wild animals, trainers often use tools such as baseball bats, chains, electric prods, and whips.

Animal Races

Horse racing and dog racing are both popular forms of entertainment. There are many abuses in both kinds of racing. Horses and dogs are often raced when they are too young. This can cause serious injuries. Sometimes dogs and horses are abused during training or during races.

Blood Sports

In these contests, animals are forced to fight against other animals or humans. Dogfights and cockfights are two examples of this type of contest. Bullfighting is also a blood sport.

CASE STUDY

Greyhound Adoption

Greyhounds were once dogs of the pharaohs of Egypt. They were bred for speed, and they love to run. Running is not a problem for greyhounds. The problem is that there are many greyhounds that are not cared for properly. Thousands of healthy young dogs and puppies are killed every year. Some are killed because they are not fast enough or because they are injured while racing. Many are killed because their owners do not want to feed them after the racing season has ended. Some are shot or poisoned. Others are abandoned and left to die or are sold to research laboratories. It is estimated that 20,000 to 25,000 greyhounds are killed every year. The lucky ones are given to greyhound rescue groups to be offered for adoption as pets. In 1992, the ASPCA, with the help of a grant from the American Greyhound Council, began to promote the adoption of retired greyhounds.

Animals in Captivity

For many people, the first time they see a wild animal is in a zoo. Many zoos today are staffed with professional people who are interested in saving wild animals from **extinction**. They carry out research that helps both captive animals and animals in the wild. The ASPCA believes that these zoos can play an important role in helping animals. However, not all animals in captivity are treated with the same respect.

Roadside Menageries

Roadside **menageries** exhibit wildlife to the public in cages alongside highways. Roadside menageries differ from zoos in that they have often been created as a way to attract people to a roadside business. The animals on display are usually not cared for properly. They live in cages that are often too small and dirty. The cages are far below the standards set by proper zoos. The animals in these menageries are often cared for by untrained staff. The ASPCA wants strict standards put in place that will eliminate menageries.

The panda is the world's most endangered species. There are about 1,000 pandas left living in China. The ASPCA believes that zoos can play an important role in helping to protect these and other endangered animals.

CASE STUDY

Marine Mammals in Captivity

The ASPCA does not believe that the welfare of whales and dolphins can be adequately provided for in captivity. These marine mammals communicate by echolocating. They send out powerful sonar signals that travel far in the oceans. In captivity, these sound waves bounce back off enclosure walls and surround dolphins and whales. This creates a stressful, noisy, confusing environment for these intelligent animals. The ASPCA is working to have laws created that would prevent the capture, breeding, and transporting of whales and dolphins.

The ASPCA feels that all places that keep captive wildlife must meet the standards set by the American Zoo and Aquarium Association. They should have properly trained staff, enclosures, and food that meet the animals' needs.

The welfare of the animals must come first in zoos and aquariums. All zoos and aquariums should work toward teaching respect and understanding of animals and their environments.

Wildlife Management

In all regions of the world, wild animals and people compete for the same space. Humans destroy wildlife **habitats** to develop cities, farmland, and recreational areas. People also hunt animals for food, furs, and sport.

Endangered Species

Many species of wildlife are endangered and threatened with extinction. Loss of habitat is a key factor. The ASPCA supports laws that protect wildlife habitats. Other threats to wildlife occur when people kill animals for their skins or other body parts. The ASPCA wants to stop the trade in animal parts, especially of endangered species.

Trapping

The trapping of wild animals for their furs causes stress and pain to the animals. If animals have to be trapped for any reason, the ASPCA promotes the use of humane traps. It also urges people to reconsider wearing fur when there are other alternatives.

Food Animals

Animals provide food for many people in the form of meat, eggs, milk, cheese, and other products. Vegetarians are people who eat plants and plant products instead of animal products. The ASPCA feels that every person must make their own decisions about which foods to eat. However, the ASPCA is concerned about the way many food animals are raised, transported, and killed. Animals used for food must be treated with respect.

Factory Farming

Today, most food animals are raised on huge **factory farms**. The animals are raised in small, crowded enclosures or pens. Most never get to go outside. Chickens are reared in tiny wire cages. The animals often become stressed and aggressive toward each other. Sometimes they bite or peck each other. To prevent pecking, factory farmers may **debeak** the chickens.

Exotic Pets

Some people keep animals that do not make appropriate pets. Owners of wild animals or exotic pets, such as lions, cougars, and monkeys, usually do not have the proper training, knowledge, or means to properly care for them. Often these animals are killed and sold to roadside menageries, or research laboratories when they get older and become dangerous. The ASPCA has **lobbied** for laws to prevent people from keeping or obtaining wild animals as pets.

Factory farms produce 98 percent of all the poultry in the United States. The chickens are kept in large sheds that often contain more than 20,000 birds.

CASE STUDY

Puppy Mills

Many of the puppies found in pet stores have come from puppy mills. Puppy mills are run by breeders who breed large numbers of puppies to sell to pet shops. The breeding dogs are often kept in poor conditions with inadequate food, shelter, or veterinary care. The poor conditions of many puppy mills can result in puppies with diseases and behavioral problems.

Not all breeders run puppy mills. There are many responsible breeders who treat their animals with care and respect. The best way to determine whether a puppy is from a puppy mill or a responsible breeder is to visit the breeder in person.

Animal shelters across North America are full of unwanted pets looking for caring homes. Still, some people turn to pet stores to provide animals.

To prevent the cruel treatment of animals in puppy mills, the ASPCA opposes the selling of dogs, cats, wild-caught birds, and wild animals through pet shops. The ASPCA is also working to have commercial dog breeding licenses removed from people who run puppy mills.

Animal Research

Animals are often used in scientific and medical research and education. Many animals have paid with their lives for scientific progress. Difficult questions have to be asked about the importance of animal welfare. The ASPCA believes that if animals must be used in research and education, their welfare should be given priority. The ASPCA has played an important role in training educators, researchers, and animal handlers who use animals in their work.

"I can only hope that our efforts to teach compassion to the young people of this country will help make this a world in which there is greater respect for human and animal life."

Larry Hawk, President of the ASPCA

Biomedical Research

The ASPCA believes that when animals are used in research, the 3R's must be applied. These are: reduce the number of animals used as much as possible; replace research animals with tissue cultures, computer programs, and other methods of testing wherever possible; and refine techniques to be sure all animals are treated humanely.

The ASPCA has lobbied for laws to protect research animals. Research facilities must now have an Internal Animal Care and Use Committee to ensure animals are treated as humanely as possible.

Quick Fact

The ASPCA has been used as a model for dozens of similar organizations across the United States, Canada, and around the world.

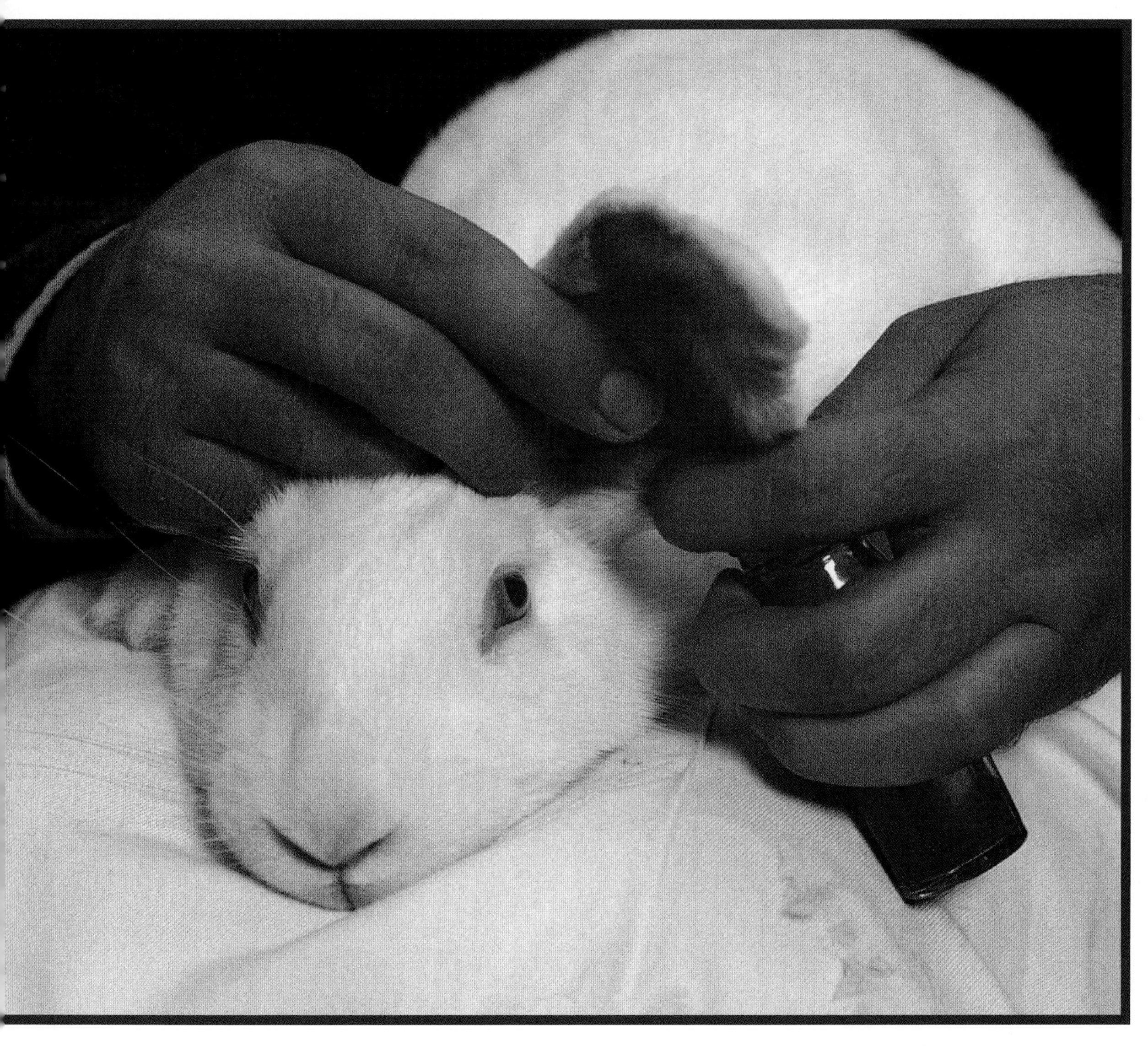

Product Testing

Many cosmetic companies have stopped testing on animals as a result of pressure from groups such as the ASPCA.

Animals are used in many ways to test products in laboratories. Some of the products are for household use such as cleaning fluids. Other products tested are cosmetics such as make-up and hair dye. Animals such as rabbits are used to test whether the products make them sick or cause pain. The ASPCA believes that all such animal testing should be stopped. The organization is lobbying to introduce new laws to prevent this kind of animal testing.

Around the World

There are more than twenty-nine countries around the world with some form of humane organization or animal welfare group. In the U.S., there are humane societies, animal shelters, and animal rescue groups in every state. The ASPCA advises and helps many of these groups through their Shelter Outreach program.

Countries with animal welfare groups are colored in yellow on this map.

CANADA
Canadian SPCA

UNITED STATES
ASPCA

GRAND CAYMAN ISLAND
The Cayman Islands Humane Society

MEXICO
Refugio Franciscano A.C.

GUATAMALA
Asociación de Amigos de los Animales

COSTA RICA
ACARA

ECUADOR
Fundación de Protección Animal

CHILE
La Protectora Punta Arenas

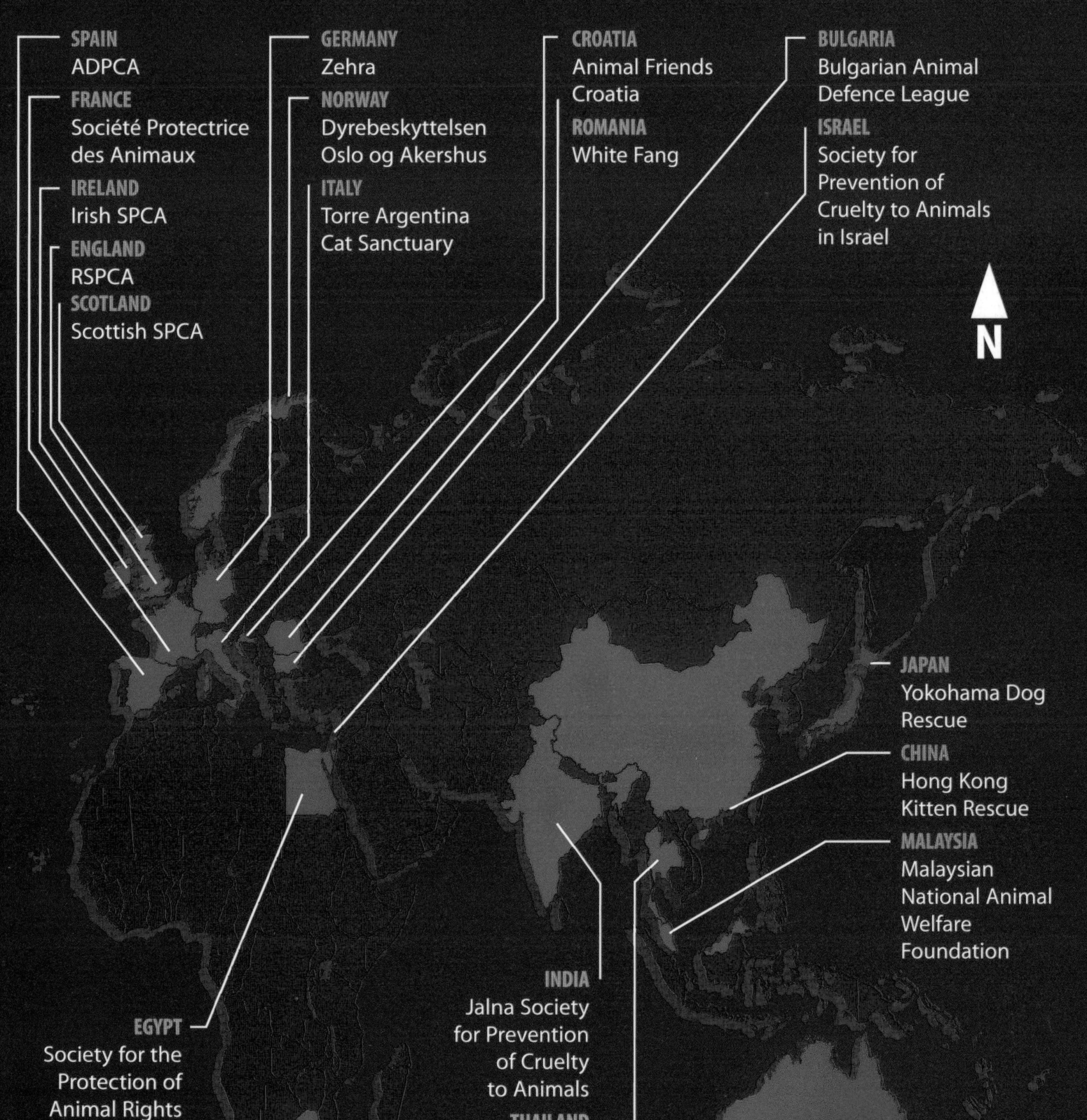
SPAIN
ADPCA
FRANCE
Société Protectrice des Animaux
IRELAND
Irish SPCA
ENGLAND
RSPCA
SCOTLAND
Scottish SPCA
GERMANY
Zehra
NORWAY
Dyrebeskyttelsen Oslo og Akershus
ITALY
Torre Argentina Cat Sanctuary
CROATIA
Animal Friends Croatia
ROMANIA
White Fang
BULGARIA
Bulgarian Animal Defence League
ISRAEL
Society for Prevention of Cruelty to Animals in Israel
N
JAPAN
Yokohama Dog Rescue
CHINA
Hong Kong Kitten Rescue
MALAYSIA
Malaysian National Animal Welfare Foundation
EGYPT
Society for the Protection of Animal Rights in Egypt
SOUTH AFRICA
SPCA of South Africa
INDIA
Jalna Society for Prevention of Cruelty to Animals
THAILAND
Thai Society for the Prevention of Cruelty to Animals
AUSTRALIA
RSPCA Australia, Inc.
NEW ZEALAND
SPCA Animal Village

Milestones

The American Society for the Prevention of Cruelty to Animals has a long and impressive record of making a difference in animals' lives. Since its start in 1866, the ASPCA has worked tirelessly to ensure that animals everywhere are treated with kindness and respect.

1867: The World's First Horse Ambulance

The ASPCA begins operating the first ambulance for horses in the world. There are many working horses in New York City. Sometimes these horses are sick, overworked, or injured. The ASPCA now has a way to rescue injured horses and move them off of the streets for care. It will be two more years before New York has an ambulance for people.

1865

Henry Bergh speaks out publicly about the importance of treating animals humanely. Many people think he is foolish, but some start to listen.

1866

Henry Bergh founds North America's first humane organization. It is called the American Society for the Prevention of Cruelty to Animals or the ASPCA. Fifty people sign the charter to start the organization.

1875: A Sling for Horses

Henry Bergh invents a special canvas sling to rescue horses trapped in mud or water. Cart horses often fell into the river or became stuck in mud while pulling heavy loads. Bergh's sling saves many horses from further injuries.

1866

The ASPCA lobbies the New York State government to make new laws to protect animals. The ASPCA is given the power to enforce the new laws.

1873

Target shooters use live pigeons as targets when they practice. The ASPCA promotes a mechanical target called a "gyro-pigeon" to be used instead of live birds.

1874

Henry Bergh rescues Mary Ellen and demands that she be given the same protection under the law as an animal. Mary Ellen is placed under the care of the court, and her foster mother is sent to prison. Soon after, Henry Bergh helps found the Society for the Prevention of Cruelty to Children (SPCC).

1894

The ASPCA takes on the job of caring for New York City's unwanted stray animals. These animals often starve or are killed in the streets. Most of the stray animals needing shelter are dogs and cats.

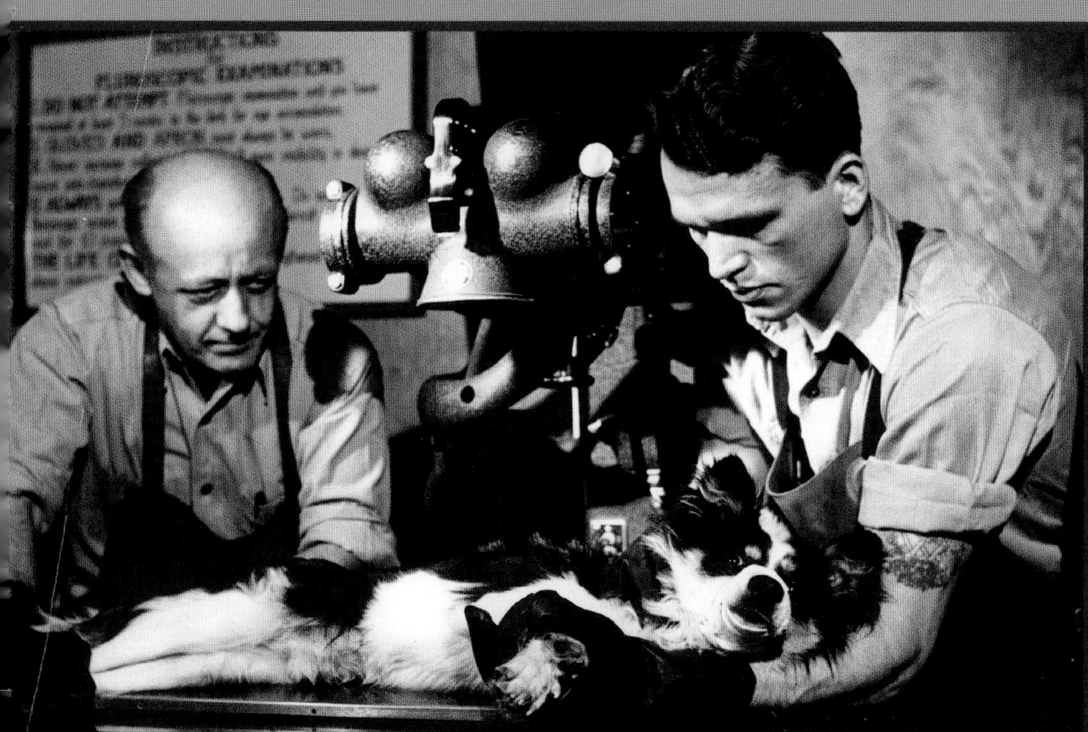

1912: First Veterinary Facility

The ASPCA offers veterinary care for the first time. Its first efforts are to provide horse owners with free treatment for their animals. The facility will later be named the Bergh Memorial Animal Hospital and become one of the largest facilities of its kind in the United States.

1902

The ASPCA uses the latest technology to help animals. The first motorized horse ambulance service is launched.

1916

The ASPCA starts to spread the word about the importance of showing kindness and respect to animals. They do this by starting a humane education program for schoolchildren.

1916

The ASPCA raises money to help care for horses that were used to carry soldiers and equipment in World War I. There are 934,000 horses that need care.

1920

The ASPCA uses **anesthetic** to help ease the pain of animal surgery. ASPCA **veterinarians** are the first to use radium to treat animals with cancer.

1925

Radio provides a new way to get people's attention. The ASPCA starts a series of radio talk shows to help spread their ideas.

1928

The humane education program is expanded. Humane educators give talks and demonstrations in school classrooms and playgrounds.

1939

The ASPCA is on guard at the New York World's Fair. It inspects 2,000 animals on exhibit at the fair to be sure they are well treated and healthy.

1942

During World War II, the ASPCA prepares for wartime emergencies. It provides courses on how to care for animals during air raids.

1944

The ASPCA recognizes the importance of a good relationship between a pet and its owner. The organization introduces classes in obedience training for dogs and their owners.

1952

For the first time, the ASPCA inspects some of the laboratories in New York that use animals for research. This is a voluntary program, but it is the first time such inspections are held in the United States.

1954

The ASPCA's animal hospital expands to better treat the many animals that need veterinary care. There is a new ward for animals with contagious diseases and new x-ray equipment. The hospital has an internship program to train veterinarians.

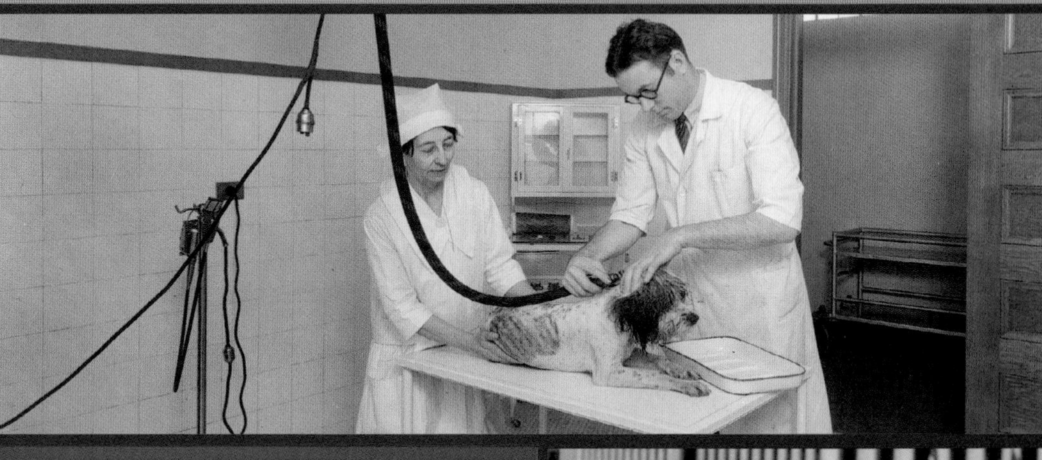

1988: New Protection for Carriage Horses

There are still many working horses in New York City. Most of these horses pull carriages that take tourists on sightseeing tours. In 1988, "Whitey," one of the carriage horses, collapses in the street from heat exhaustion. This precipitates the first change to the carriage horse law in many years. ASPCA agents now carry special thermometers in the summer to measure the temperature around the carriage horses. It is against the law to work carriage horses when the temperature is above 90° Fahrenheit. ASPCA agents also make sure the horses are given water to drink.

1958

The ASPCA opens the Animalport at Kennedy International Airport. This is where animals arriving in or departing the United States by airplane are inspected and cared for.

1961

ASPCA veterinarians save a dog's life by performing open-heart surgery. It is the first time open-heart surgery is performed on a dog.

1976

The Bergh bandage is developed by Dr. Gordon Robinson for treating animals at the ASPCA. The bandage works very well and is now used across the U.S.

1985

The ASPCA has a new Government Affairs office in Washington, DC. From here, it lobbies the government for new and improved animal protection laws. The organization also ensures that existing laws benefit animals.

1996: The National Animal Poison Center

Operation of the National Animal Poison Center is taken over by the ASPCA. The center was previously run by the University of Illinois College of Veterinary Medicine. The center gives veterinary advice and help in cases of animal poisoning. Its telephone service is available 24 hours a day, every day of the year.

2000: Partnership with Petfinder

The ASPCA's National Shelter Outreach program partners with Petfinder.com. Petfinder works as a cyber shelter, and matches homeless pets with people looking for pets. Within two years, nearly 4,000 animal shelters and rescue groups are participating in the Petfinder program. In 2000 and 2001, more than 1.2 million animals find homes through the Petfinder program.

1985

The ASPCA works to have the Animal Welfare Act revised. It wants to ensure protection of research animals. If intelligent and active animals, such as dogs and monkeys, are used in research, their special needs must be met.

1992

The ASPCA urges people to adopt retired greyhounds. Many of these dogs are destroyed after a few years of racing. The ASPCA helps groups across the U.S. who work to save these dogs by having them adopted as pets.

1993

The ASPCA joins ten other humane groups to begin a National Council on Pet Population Study and Policy. For the first time, researchers know how many and what kinds of animals are in shelters in the United States.

1994

The ASPCA helps pass the New York State Animal Experimentation Bill. This gives students who do not want to dissect animals in science classes a choice. They can choose to do a different project worth the same amount of marks.

1996

The Henry Bergh Memorial Hospital starts using the Care-A-Van. This is a mobile spay and neuter clinic. It travels to inner-city neighborhoods to help control pet overpopulation.

1998

A law is passed that allows residents of public housing to keep pets.

2000

The Chimp Sanctuary Bill and Safe Air Travel for Animals Act are passed.

2000

The Henry Bergh Children's Book Award is launched.

Current Initiatives

In the new millennium, the ASPCA renewed its commitment to ease the stress and abuse of working animals, farm animals, pets, and wildlife across the country. To continue its mission, the ASPCA has initiated new laws and programs.

"The explosion of computer and Internet technology has allowed us to ... impact the lives of more animals faster than ever before."
ASPCA

New Laws

The ASPCA's Government Affairs Department continues to work for the welfare of animals. In 2000, the Safe Air Travel for Animals Act was passed. The Act requires the reporting of animals injured, lost, or killed during air transport. It also requires airline and baggage personnel to be trained in the proper care and handling of animals. The ASPCA also worked to pass the Chimp Sanctuary Bill in 2000. This bill established **sanctuaries** for chimpanzees who are no longer used in research.

Other laws that the ASPCA has recently initiated include a bill that allows retired military dogs to be adopted instead of being put down. A law banning the marketing of cat and dog fur was also recently passed.

Increasing Animal Adoptions

The ASPCA uses the Internet to help more animals than ever before find new homes. The ASPCA's National Shelter Outreach program partnered with Petfinder.com in 2000. Petfinder is a cyber shelter that helps match homeless pets with people looking for pets. People looking for pets type in their zip code and the kind of pet they want. Photos of animals available for adoption in their area appear on the computer screen. The program has proven to be a successful way to save animals' lives and give them a fresh start in a caring home.

Every year, 3 to 4 million unwanted cats and kittens have to be put down because homes cannot be found for them.

CASE STUDY

DOGS' BEST FRIEND

Wyatt's Ruff Rescues is one of the smallest rescue groups on Petfinder.com. Ruff Rescue is a one-person rescue operation. It was formed by Wyatt Tyler in 2000 to save the lives of dogs at the local animal shelter in Wauchula, Florida. Wyatt was only 9 years old when he started using his Christmas, report card, and tooth fairy money to adopt dogs whose time was up at the animal shelter. Dr. Ross Hendry, a local veterinarian, agreed to provide free spaying and neutering to Wyatt's dogs. He also gives them low-cost rabies shots. Wyatt promotes his dogs on Petfinder.com to find new homes for them. Until he finds them homes, Wyatt has to feed, water, train, and care for his rescued dogs. Wyatt loves looking after dogs. He is now 10 years old and with the help of Dr. Hendry and Animal Control, he has saved the lives of ninety-eight dogs and five cats. Donations are now coming from schoolmates and many other people. Wyatt uses donations to help adopt more dogs, pay for food and vet bills, and to build kennels. Wyatt Tyler is proof that one person can make a big difference.

Intervention for Animal Abusers

In the late 1990s, the ASPCA Counselling Services began an intervention program for convicted animal abusers. It was the first such program in the United States. The program gives the justice system a tool for dealing with convicted animal abusers. The program has personality profiles of abusers that can be used to recommend treatment. The ASPCA Counselling Services is involved with the New York Family Vision program. This program looks at links between animal abuse and family violence.

New Directions in Veterinary Care

The Bergh Memorial Animal Hospital has provided care for injured, abused, and sick animals since 1912. Today, it treats over 25,000 animals a year. In 2000, its intensive care unit expanded to handle up to forty critically sick animals. A new veterinary staff position was created to focus on animal abuse cases. This allows the ASPCA to better recognize and prosecute abuse cases that come to the hospital. New surgical procedures for spaying and neutering animals allow animal patients to have shorter hospital stays. This means more animals can receive these operations.

Reaching Children

The education of children is still an important goal of the ASPCA. To help achieve this, the ASPCA launched the Henry Bergh Children's Book Award in 2000. This award honors books that promote care, compassion, and respect for all living creatures.

In 2001, the ASPCA launched a children's Web site. The site is Animaland.org and it contains information, games, and stories about animals.

Quick Fact

The ASPCA National Shelter Outreach department has helped more than 1.2 million animals find homes since 2000.

CASE STUDY

SAVING ANIMALS AT GROUND ZERO

"A lot of the animals were treated for dehydration, respiratory illness, and shock. But most were surprisingly healthy and just all excited to see their owners of course."
Brigid Fitzgerald, ASPCA

On September 11, 2001, the ASPCA was called upon to provide animal disaster relief on a larger scale than ever before. The terrorist attacks on the World Trade Center in New York City caused panic and confusion. Firefighters, police officers, emergency crews, and volunteers sprang into action to help human survivors of the attacks.

Many people were also concerned about helping animal survivors. The ASPCA set up a Rescue Center to help animals that were injured, orphaned, lost, or trapped during the attacks. Mobile veterinary units were set up to provide emergency care. A pet rescue hotline was established. People could telephone in reports of stranded or lost animals. The ASPCA's Humane Law Enforcement officers took people evacuated from apartments back to get their pets. Any animals whose owners were missing were taken to safety. Many of these animals were later reunited with their owners.

The ASPCA provided veterinary care to 300 animals rescued or removed from apartments following the attacks. Volunteer help and donations of food, money, and veterinary supplies came from across the United States and Canada.

Take Action!

Become an active and responsible citizen by taking action in your community. Participating in local projects can have far-reaching results. You do not have to go overseas to get involved. You can do service projects no matter where you live. In fact, young people are helping out every day. Some help support overseas projects. Others volunteer for projects in their home communities. Here are some suggestions:

Volunteer some of your time and talents at your local animal shelter.

Learn how to make safe cat toys and dog treats to brighten up the lives of animals at the shelter. Visit the Humane Education part of the ASPCA Web site to learn more.

Write a letter or article to your school or community newspaper. Tell people how important it is to have their pets spayed or neutered.

Take a dog for a walk, or groom a cat or dog. Show others how to be a responsible pet owner by being one yourself.

Find out what your local shelter needs most and help raise money for it. You and your friends can hold a garage or bake sale, walk neighbors' dogs, babysit, cut lawns, or recycle bottles or newspapers to raise money. You can probably think of other ways to raise money.

Encourage friends or neighbors who are looking for a pet to visit a shelter or to look at the animals on the Petfinder and ASPCA Web sites.

Become a member of the ASPCA and other groups that promote animal welfare and conservation.

Where to Write

International

RSPCA
Enquiries Service
Wilberforce Way
Southwater
West Sussex
RH13 9RS
UK

World Animal Net
24 Barleyfields
Didcot, Oxon
OX11 OBJ
UK

Animal Aid
The Old Chapel
Bradford Street
Tonbridge, Kent
TN9 1AW
UK

United States

ASPCA
424 E. 92nd Street
New York, NY
10128-6804

ASPCA Animal Poison Control Center
1717 South Philo Road
Suite 36
Urbana, IL
61802-6044

PETA
501 Front Street
Norfolk, VA
23510-1009

Canada

Canadian SPCA
5215 Jean-Talon W.
Montreal, QC
H4P 1X4

The Animal Rescue Foundation
PO Box 34160
Calgary, AB
T3C 1S2

In the Classroom

EXERCISE ONE:

Make Your Own Brochure

Organizations such as the ASPCA use brochures to inform the public about their activities. To make your own ASPCA brochure, you will need:

- paper
- ruler
- pencil
- color pens or markers

1. Using your ruler as a guide, fold a piece of paper into three equal parts. Your brochure should now have a cover page, a back page, and inside pages.
2. Using your color markers, design a cover page for your brochure. Make sure you include a title.
3. Divide the inside pages into sections. Use the following questions as a guide.
   - What is the organization?
   - How did it get started?
   - Who started it?
   - Who does it help?
4. Using the information found in this book, summarize in point form the key ideas for each topic. Add photographs or illustrations.
5. On the back page, write down the address and contact information for the ASPCA.
6. Photocopy your brochure and give copies to your friends, family, and classmates.

EXERCISE TWO:

Send a Letter to Your Congressperson

To express concern about a particular issue, you can write a letter to your member of congress. This can be an effective way to make the government aware of issues that need its attention. To write a letter, all you need is a pen and paper or a computer.

1. Find out the name and address of your congressperson by contacting your local librarian. You can also search the Internet.
2. Write your name, address, and phone number at the top of the letter.
3. When addressing your letter, use the congressperson's official title.
4. Outline your concerns in the body of the letter. Share any personal experiences you may have that relate to your concerns. Use information found in this book to strengthen your concerns.
5. Request a reply to your letter. This ensures that your letter has been read.
6. Ask your friends and family to write their own letters.

Further Reading

Goodman, Susan E. *Animal Rescue: The Best Job There Is*. New York: Simon and Schuster Books for Young Readers, 2000.

Loeper, John. *Crusade for Kindness*. New York: Atheneum Books for Young Readers, 1991.

Mandel, Gerry and William Rubel (eds.) *Animal Stories by Young Writers*. Berkeley, CA: Tricycle Press, 2000.

Sirch, Ann Willow. *Careers With Animals*. Golden, CO: Fulcrum Publishing, 2000.

Sutton, John G. *Animals Make You Feel Better*. Shaftesbury, Dorset: Element Children's Books, Inc., 1998.

Web Sites

Animaland
www.animaland.org
Animaland is the ASPCA's Web site for children. The site contains games, stories, and trivia. It also has pet care tips and a career center for animal lovers.

ASPCA Web site
www.aspca.org
The official Web site of the ASPCA provides visitors with a history of the organization and insight into its operations. There are links to ASPCA initiatives such as the Poison Control Center.

Petfinder Cyber Shelter
www.petfinder.org
Petfinder is an online database of more than 60,000 adoptable animals. The site is connected with more than 4,000 animal shelters across North America. Users can enter their zip code to find pictures of animals available for adoption in their neighborhood.

Glossary

anesthetic: a type of drug that blocks feelings of pain
debeak: to remove a bird's beak
domestic: animals like cows and dogs that have been bred for use by humans
endangered species: species that are in danger of disappearing
extinction: the complete disappearance of a species
factory farms: large farms where animals are raised for food in close confinement
habitats: areas where animals live and where they find their food, water, and shelter
humane: behaving in a way that promotes kindness and caring toward all beings
lobbied: campaigned for a particular issue or law
menageries: collections of animals put on display
neuter: an operation done by a veterinarian to prevent a male animal from breeding
sanctuaries: safe homes for animals
spay: an operation done by a veterinarian to prevent a female animal from breeding
veterinarians: animal doctors

Index